Time Passages

There is a story about a very young student who was asked to color a picture of a flower. Being only 5 years old he colored a black flower with purple leaves. His teacher took his picture away and gave him another copy to color. She told him that flowers are not black. Flowers are RED and leaves are Green. He dutifully colored his flower as he was told and every flower he colored after that was red with green leaves, and soon he learned to wait for the teacher to tell him what to do and how to do it. Several years later another teacher told her class to color a flower. The little boy sat and didn't start coloring. When the teacher asked why he wasn't coloring his flower he said he was waiting for her to tell him what color it should be. "Make it any color you want," said the teacher, "use your imagination." The boy turned to his picture and colored his flower red and the leaves green.

With this book, please do not be like the little boy. If you have lost the reckless creative imagination of your youth, then please find it again here. Be as creative and imaginative as you want to be. The pictures on the covers are there to spur ideas and NOT to show what pictures should look like.

You will find that some pictures have more detail than others, some lend themselves to the addition of details more than others, some are greyscale and some are wire form. Please enjoy them all to the extent that your imagination allows.

COMPLETED PICTURES WILL NOT BE MARKED, RATED, COMPARED, ANALYSED or JUDGED!

Enjoy!

See a selection of recommended art supplies on the last page

HINT: When beginning with a wire form image, start by drawing a heavier line around the area which you intend to fill with color.

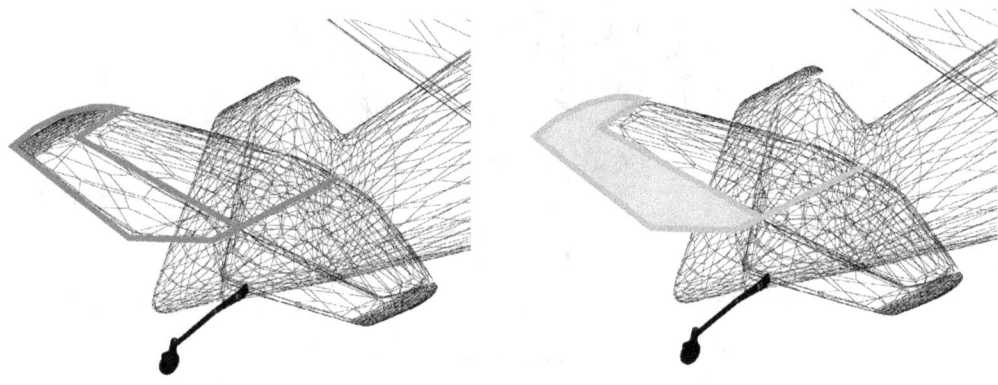

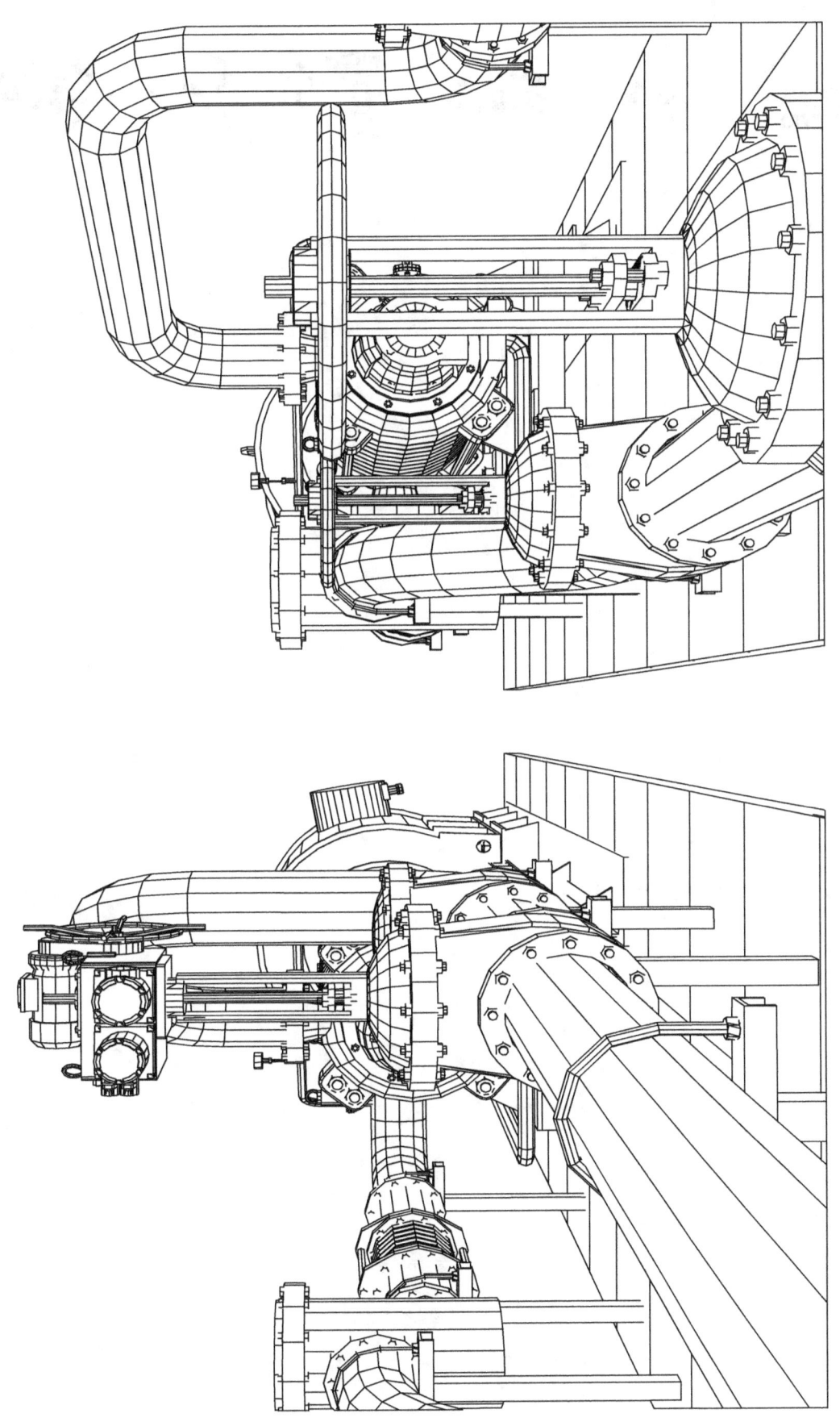

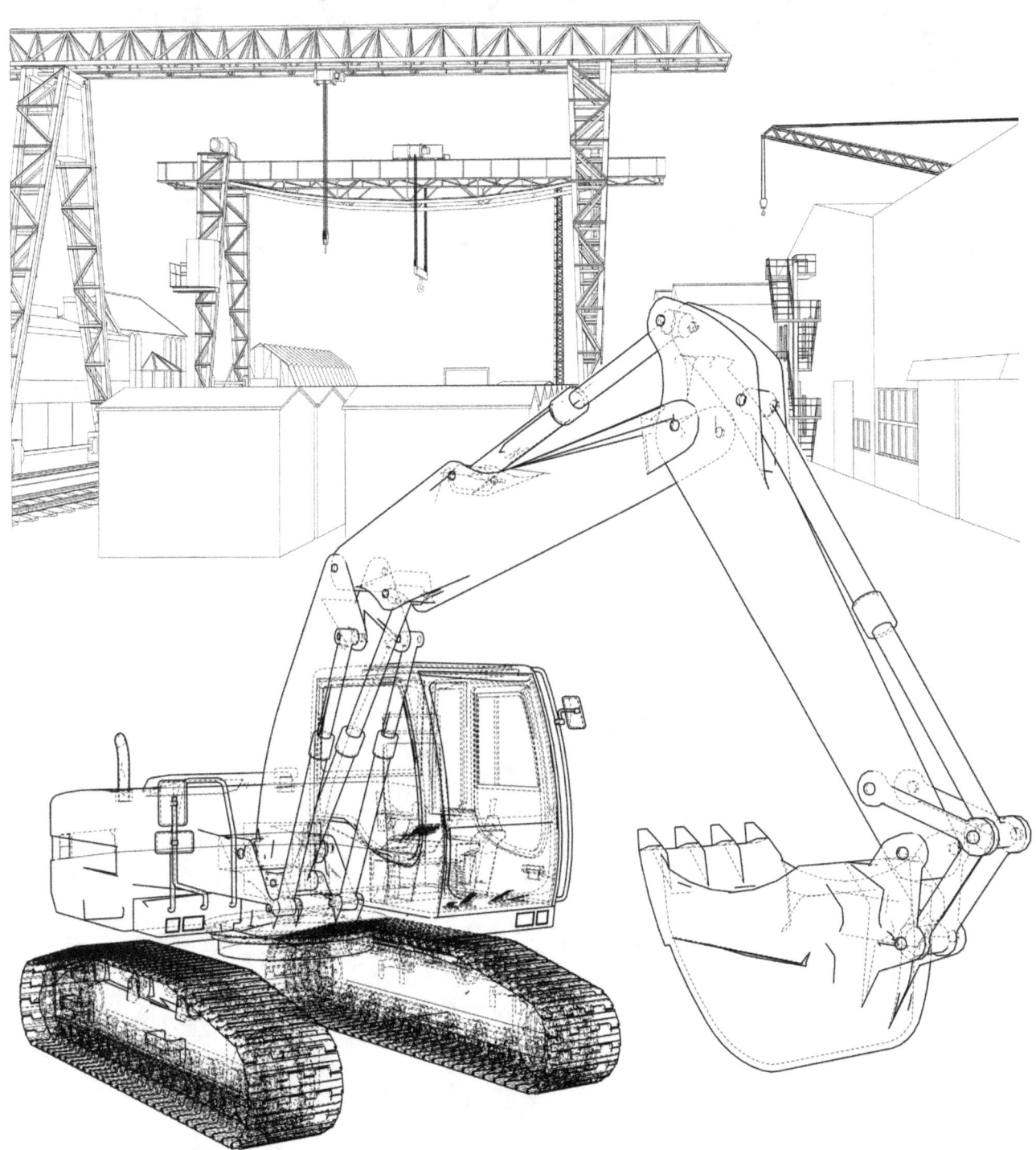

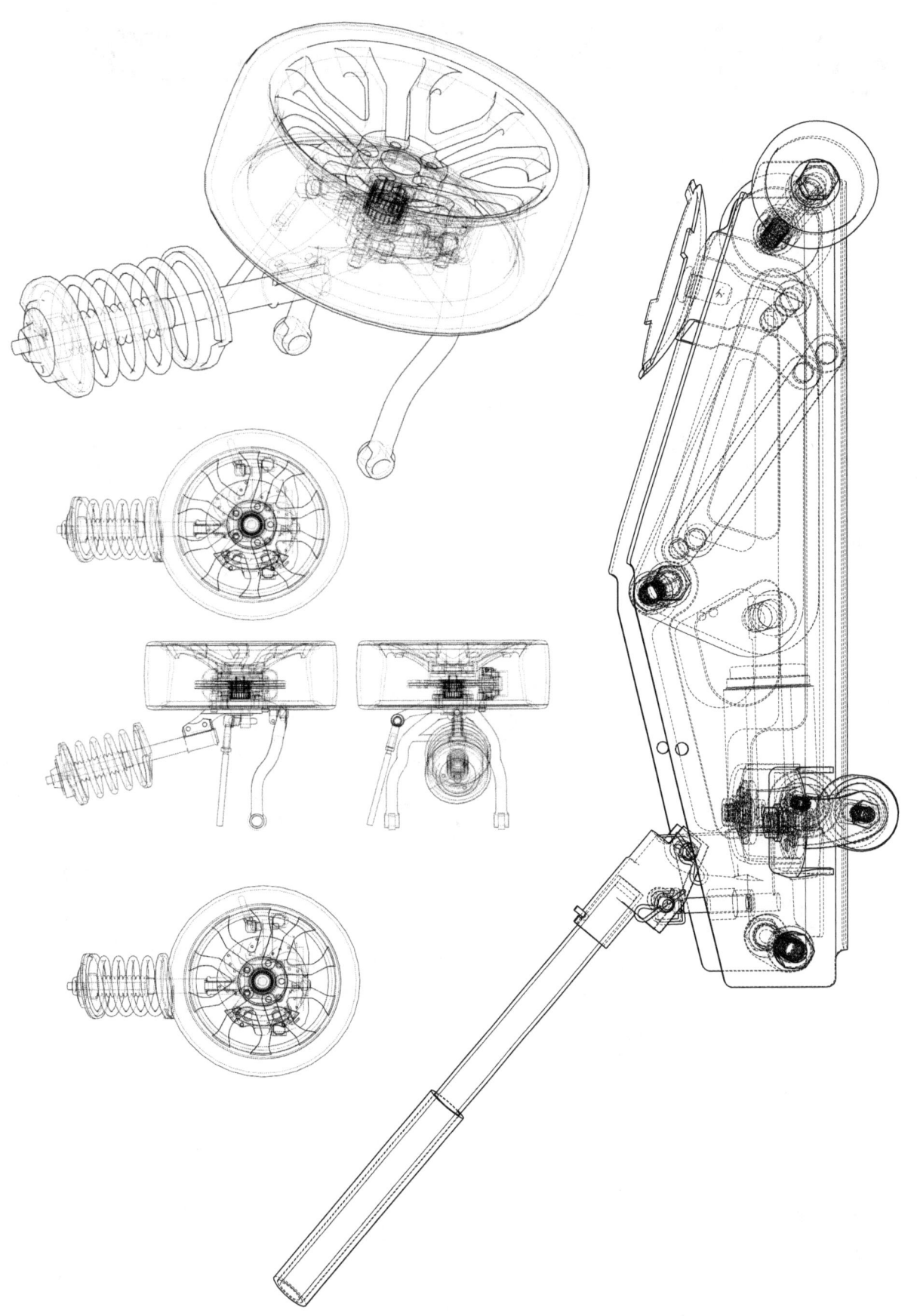

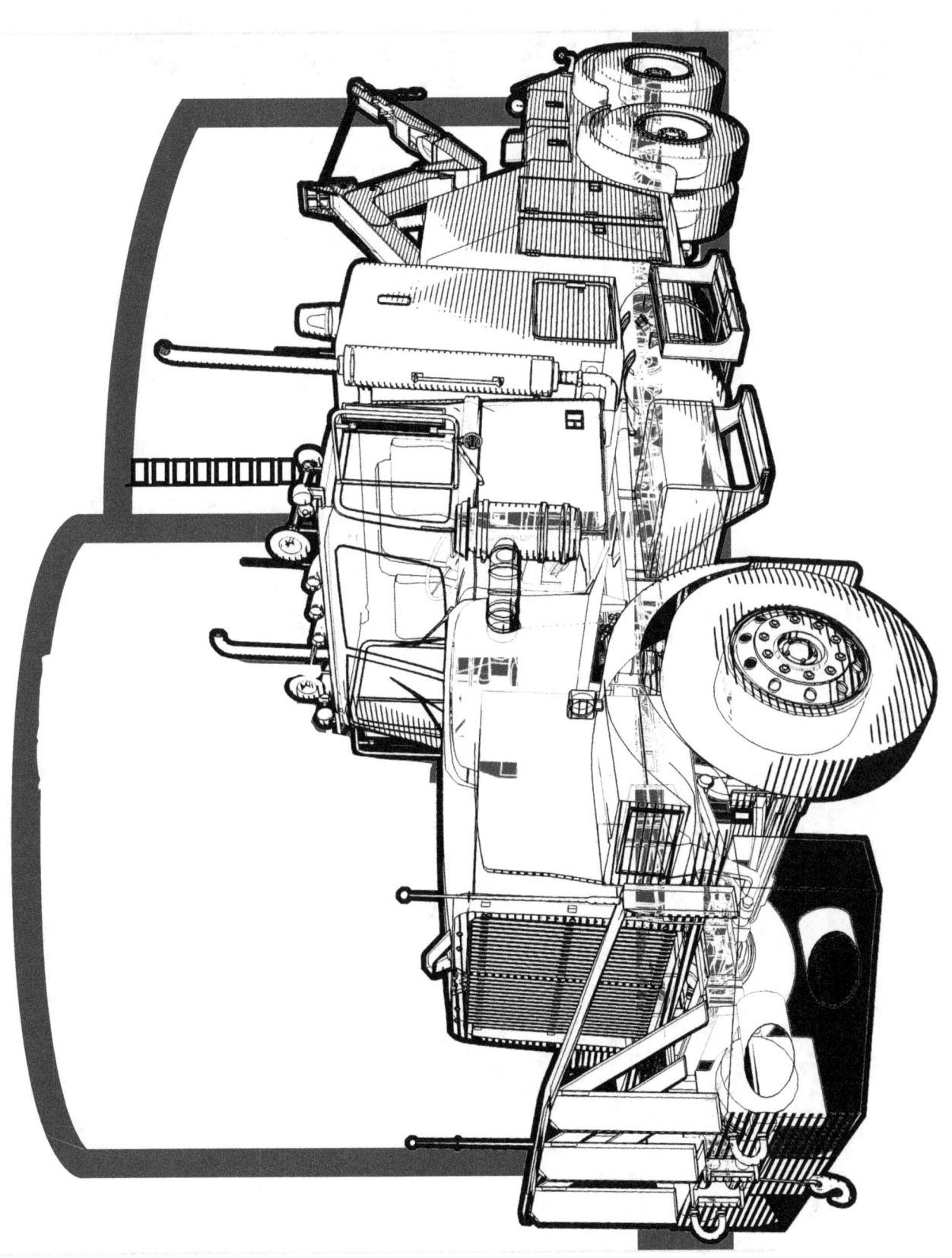

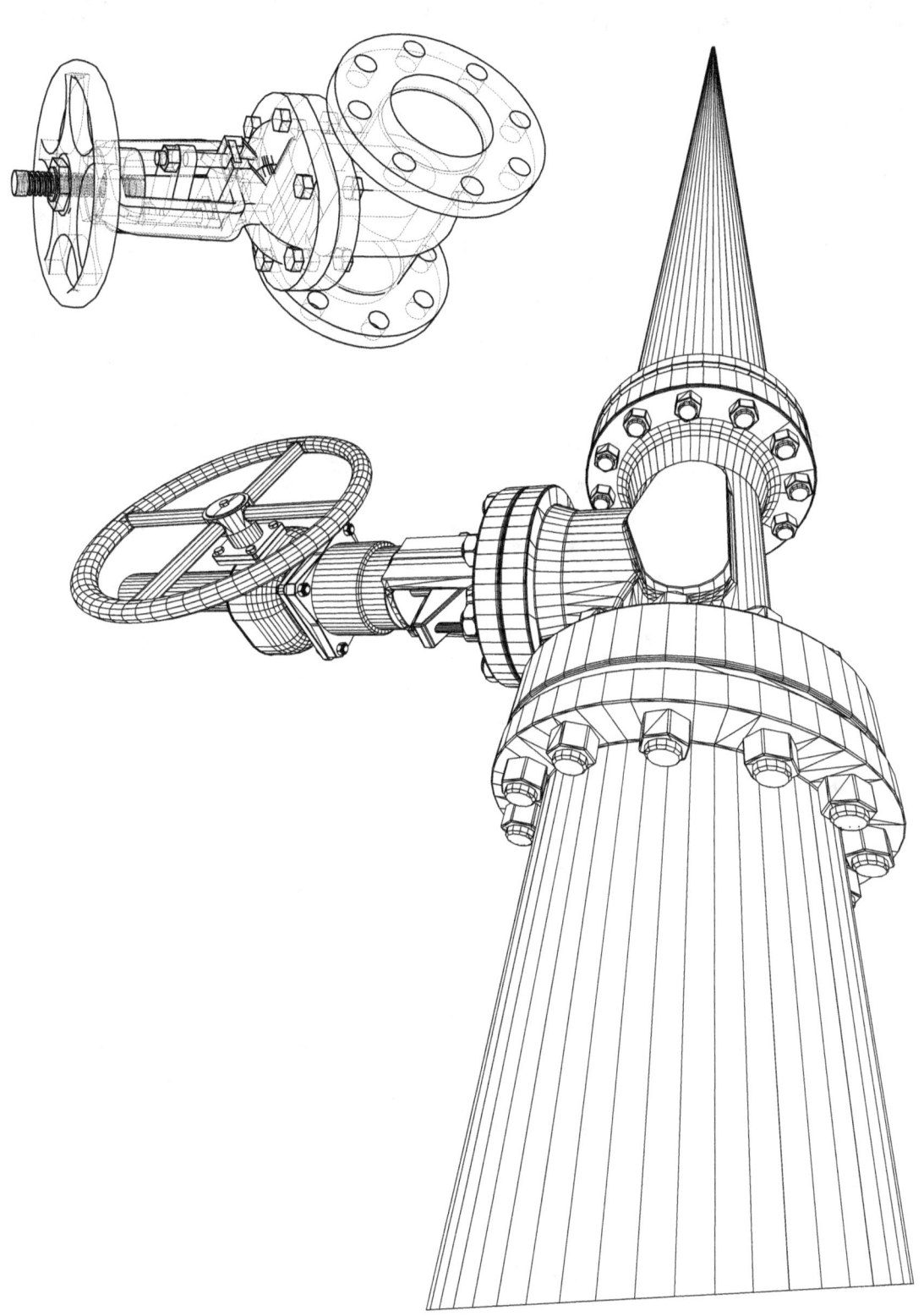

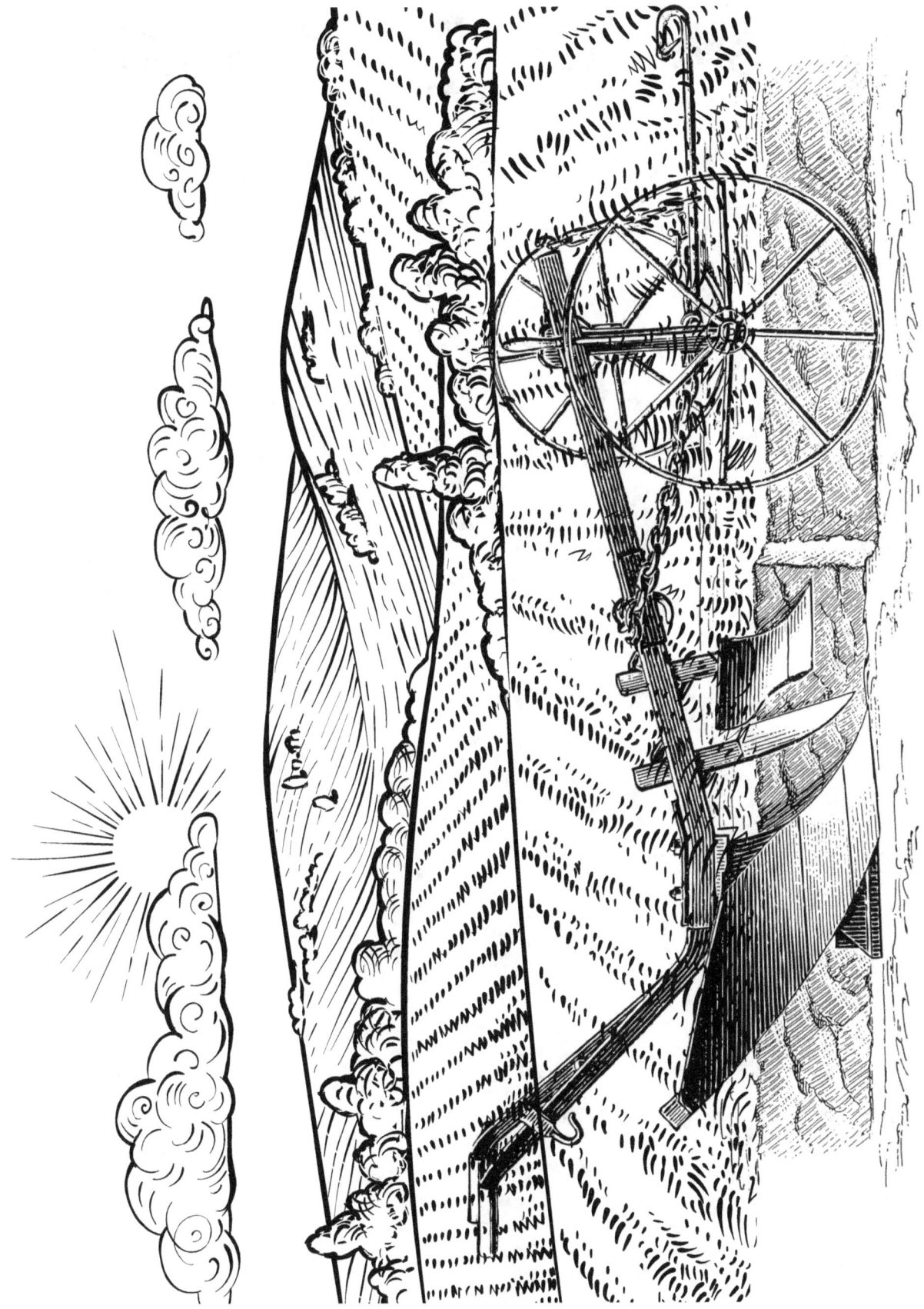

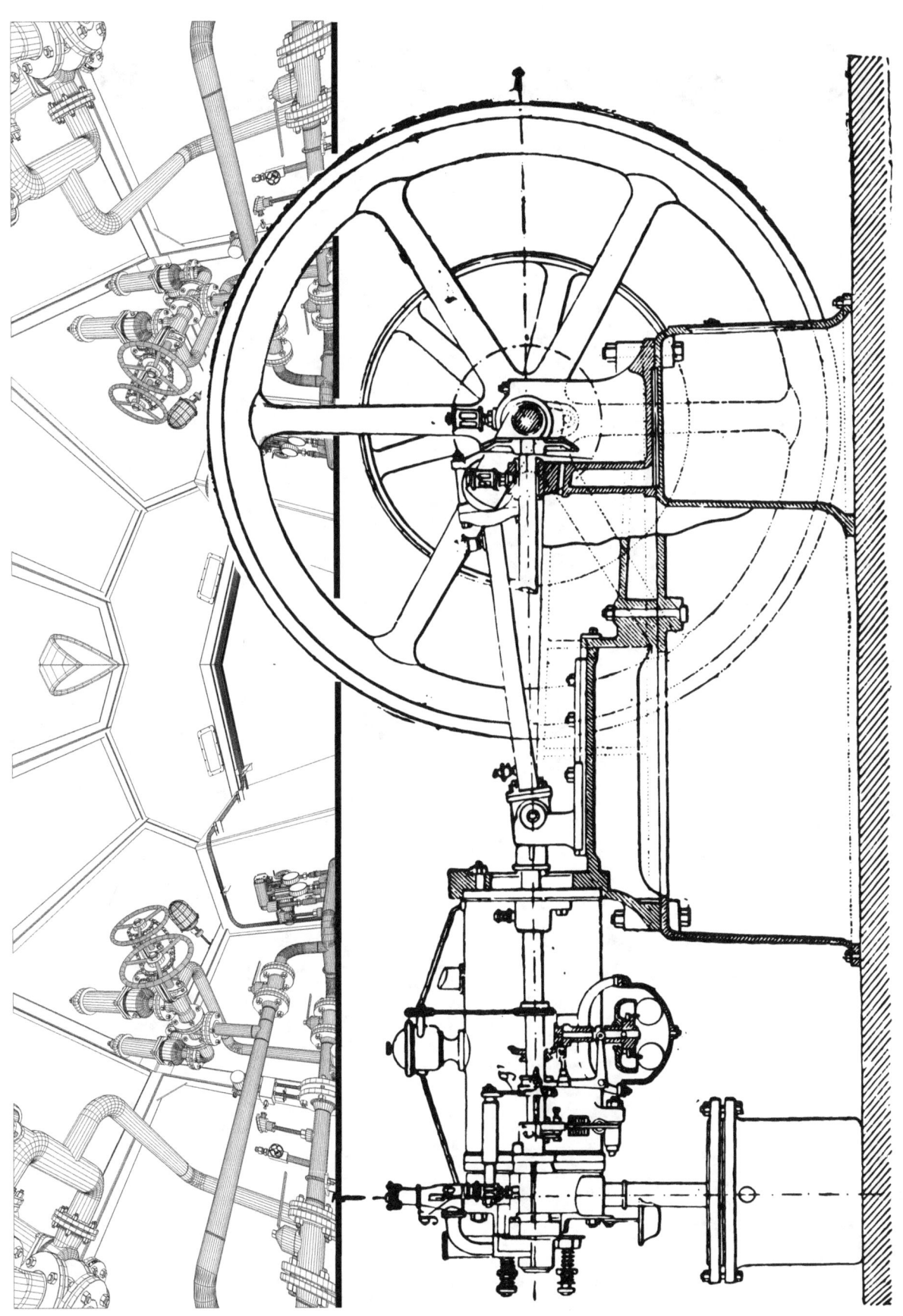

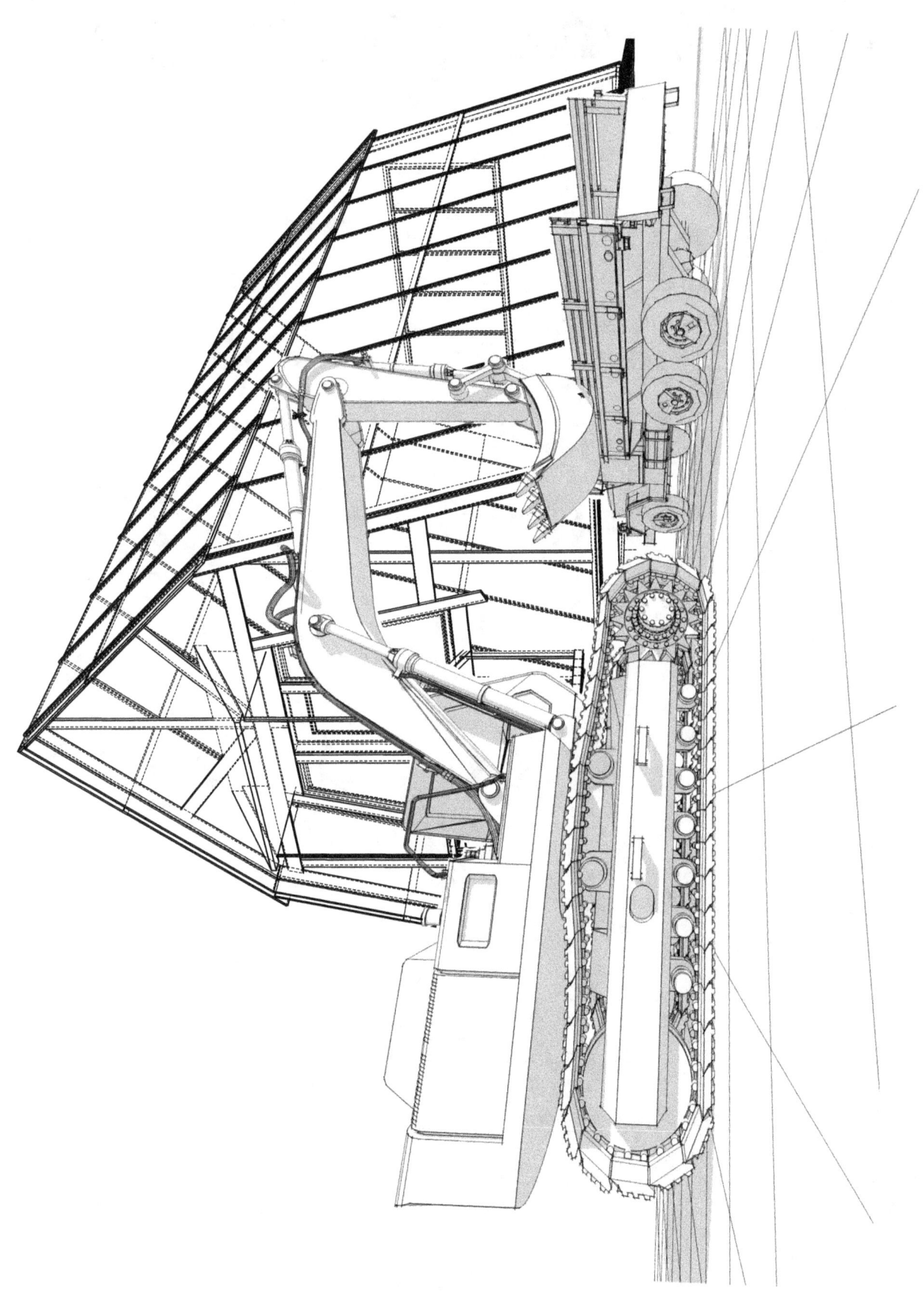

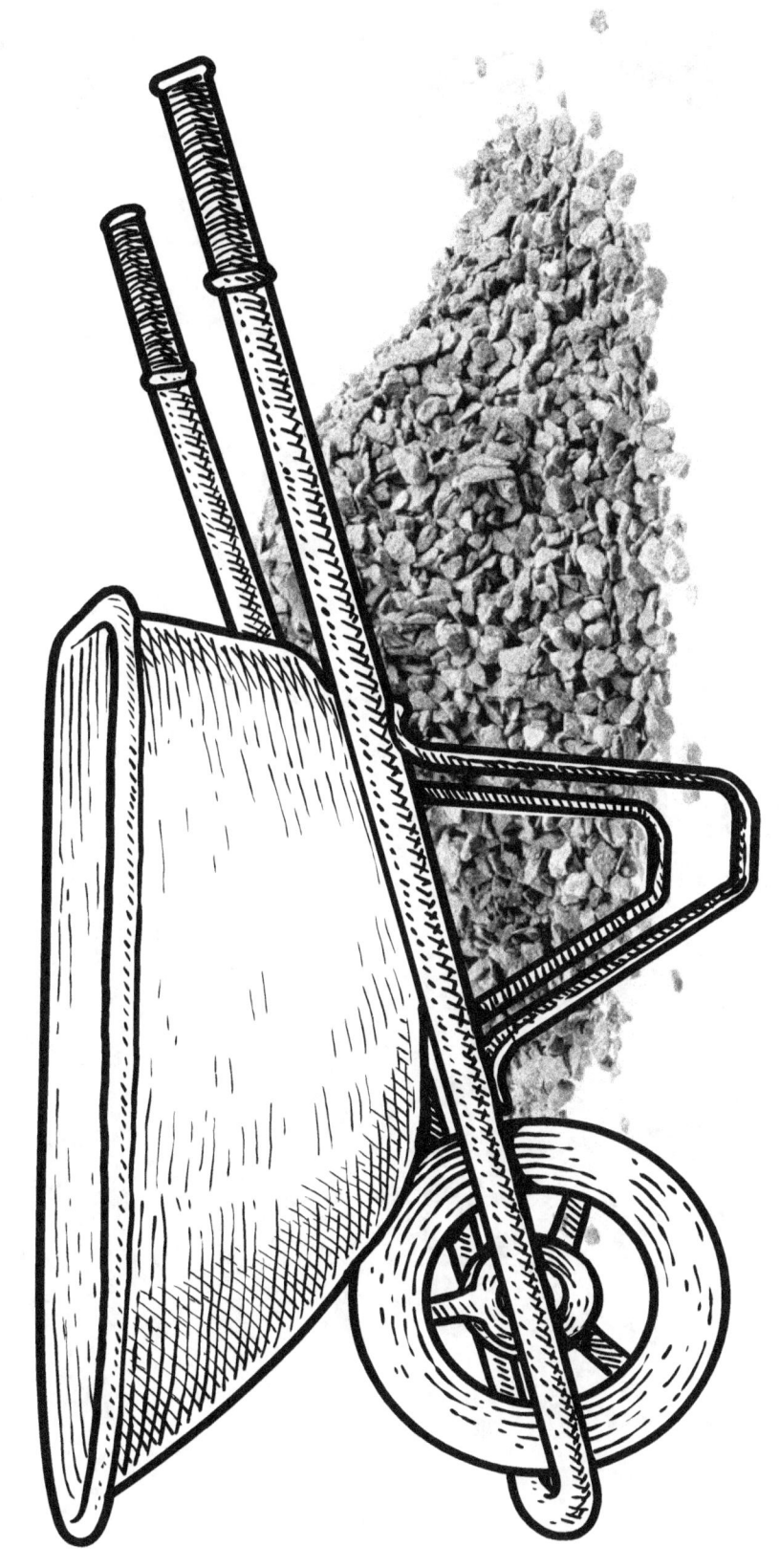

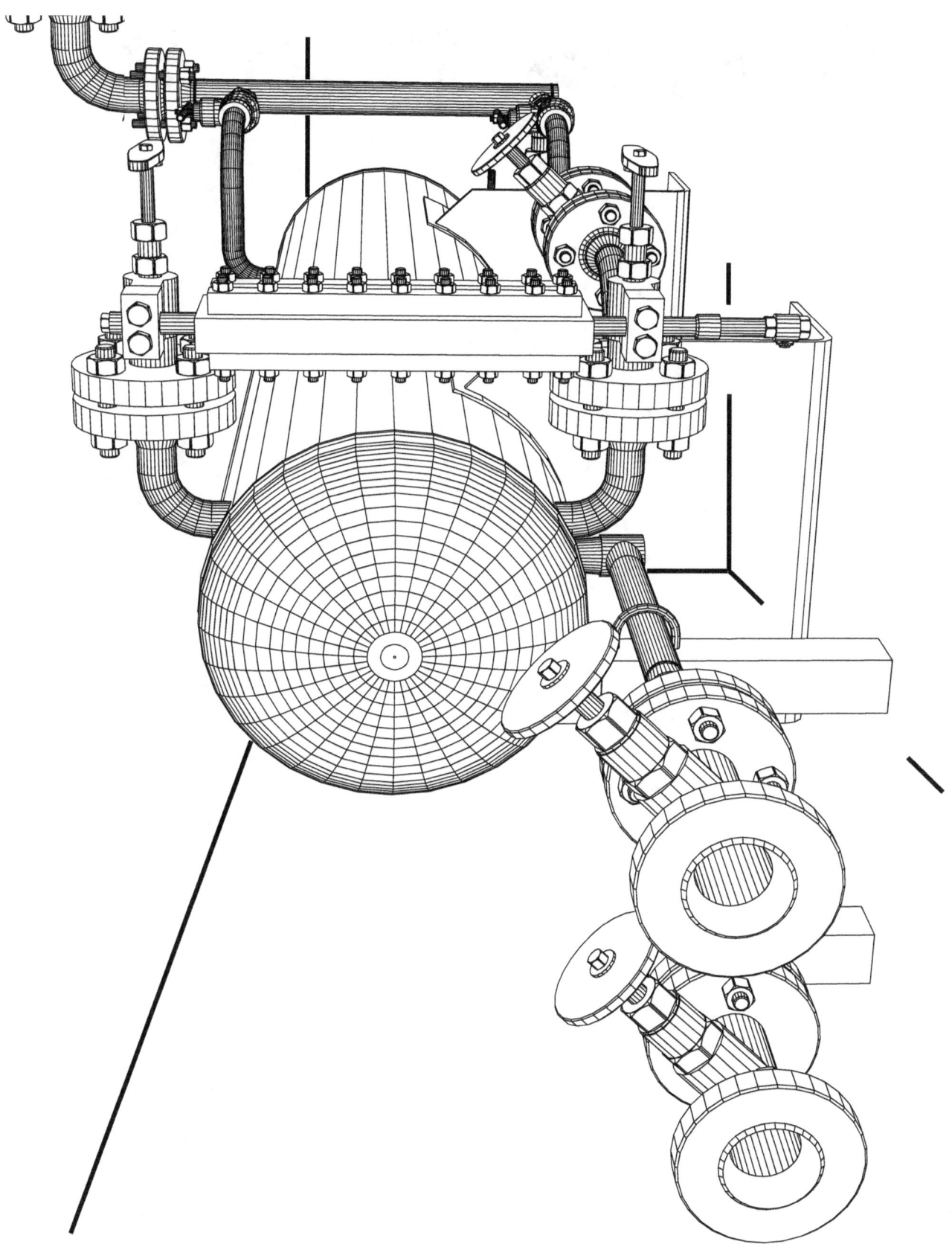

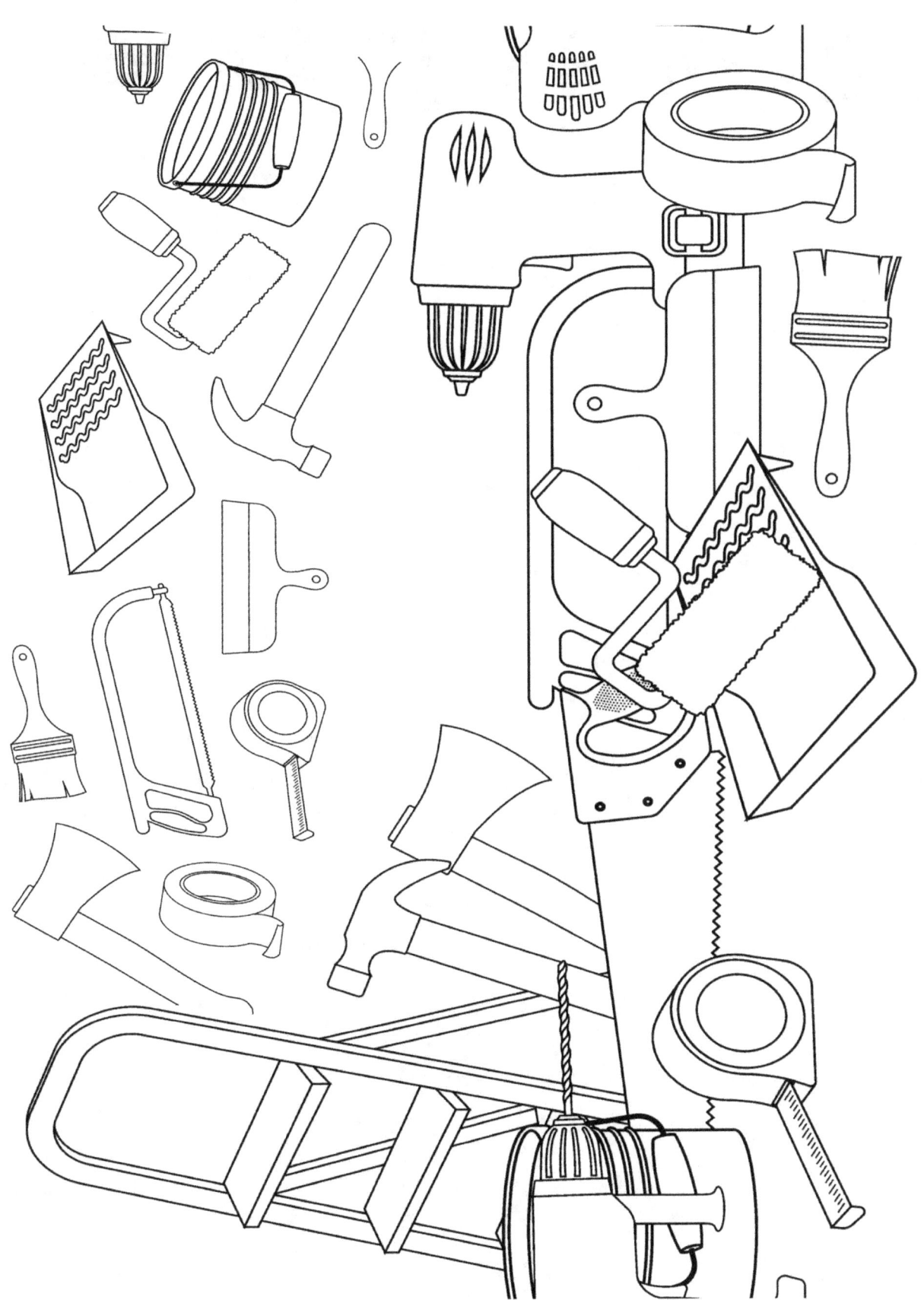

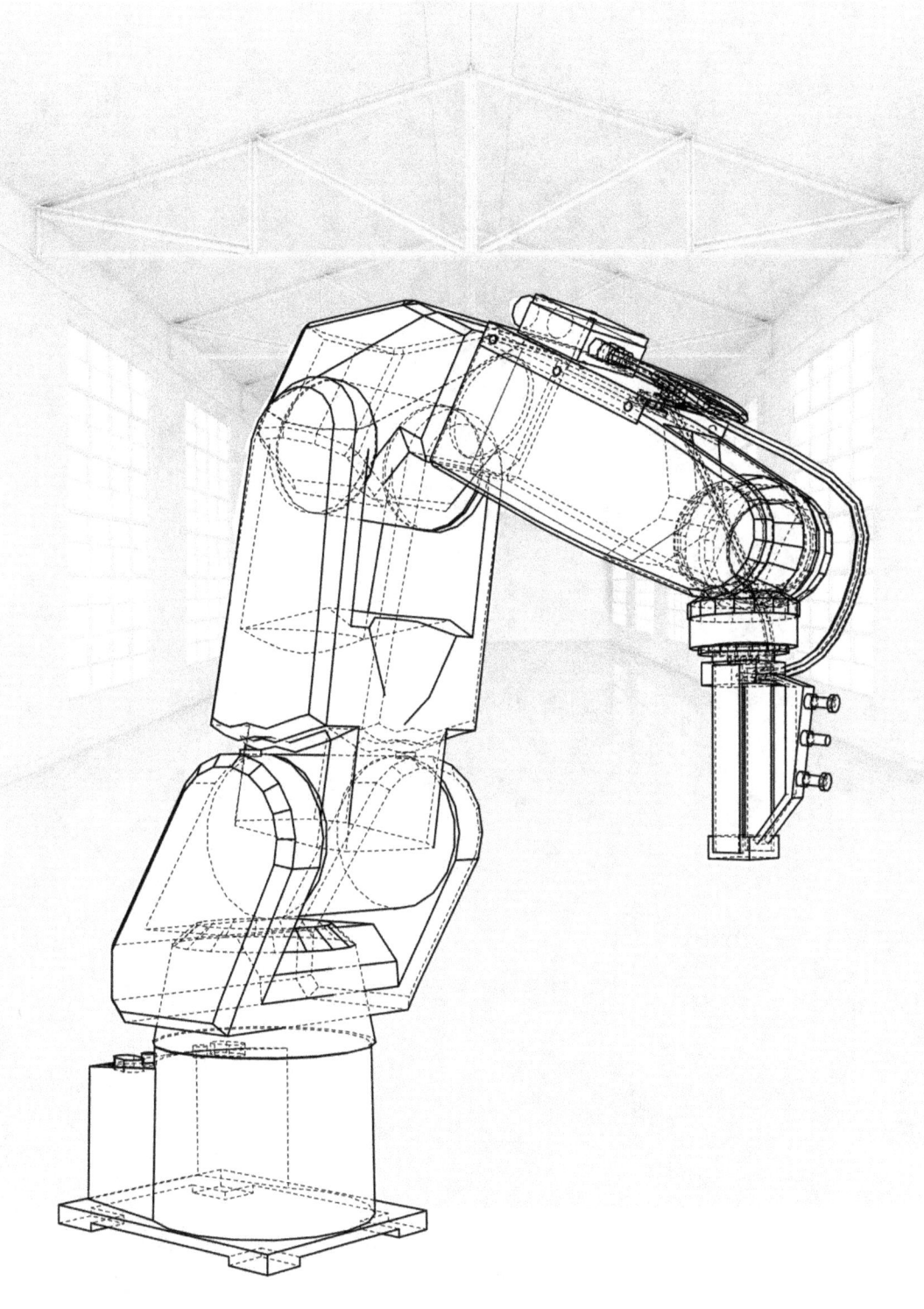

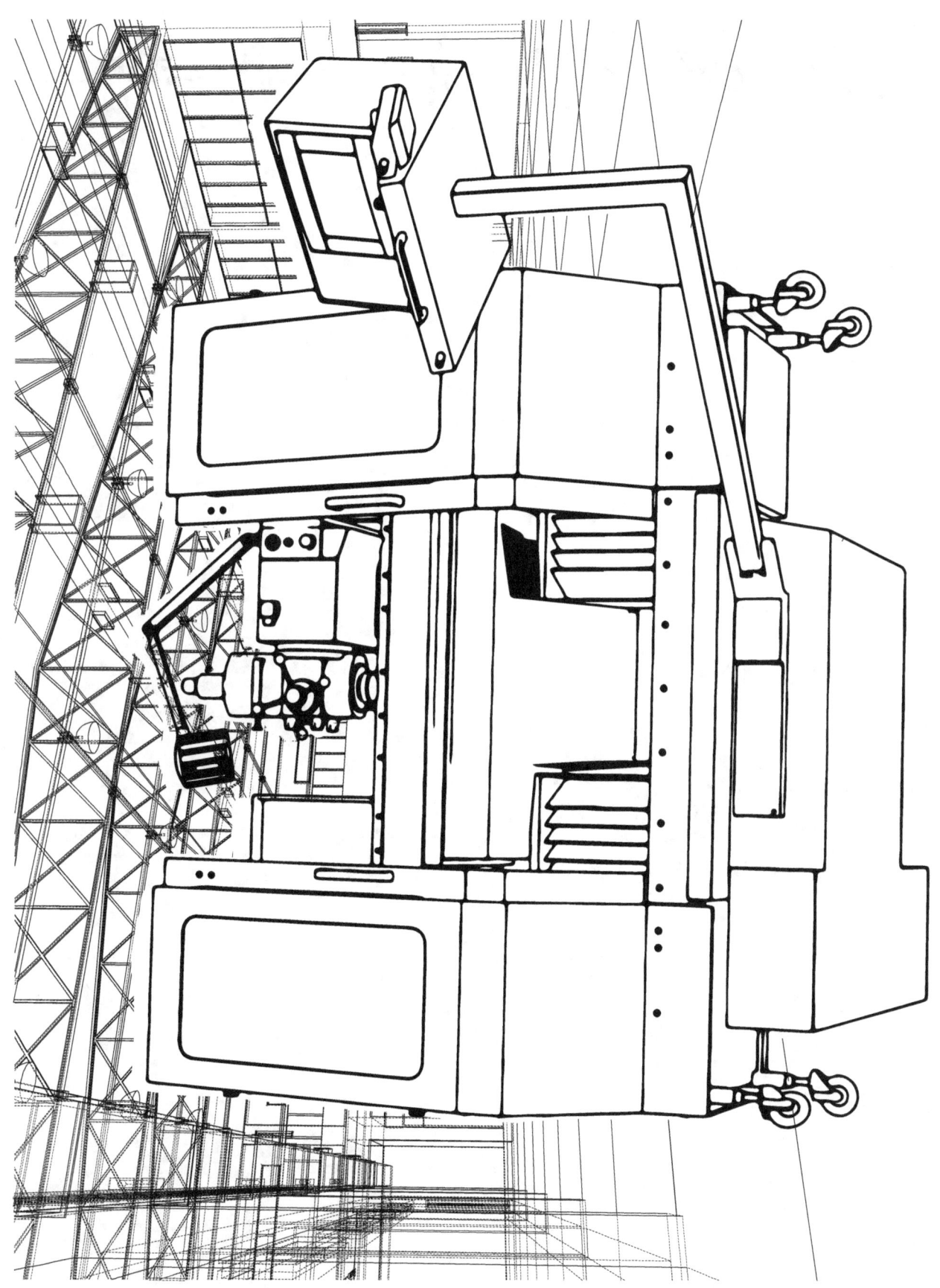

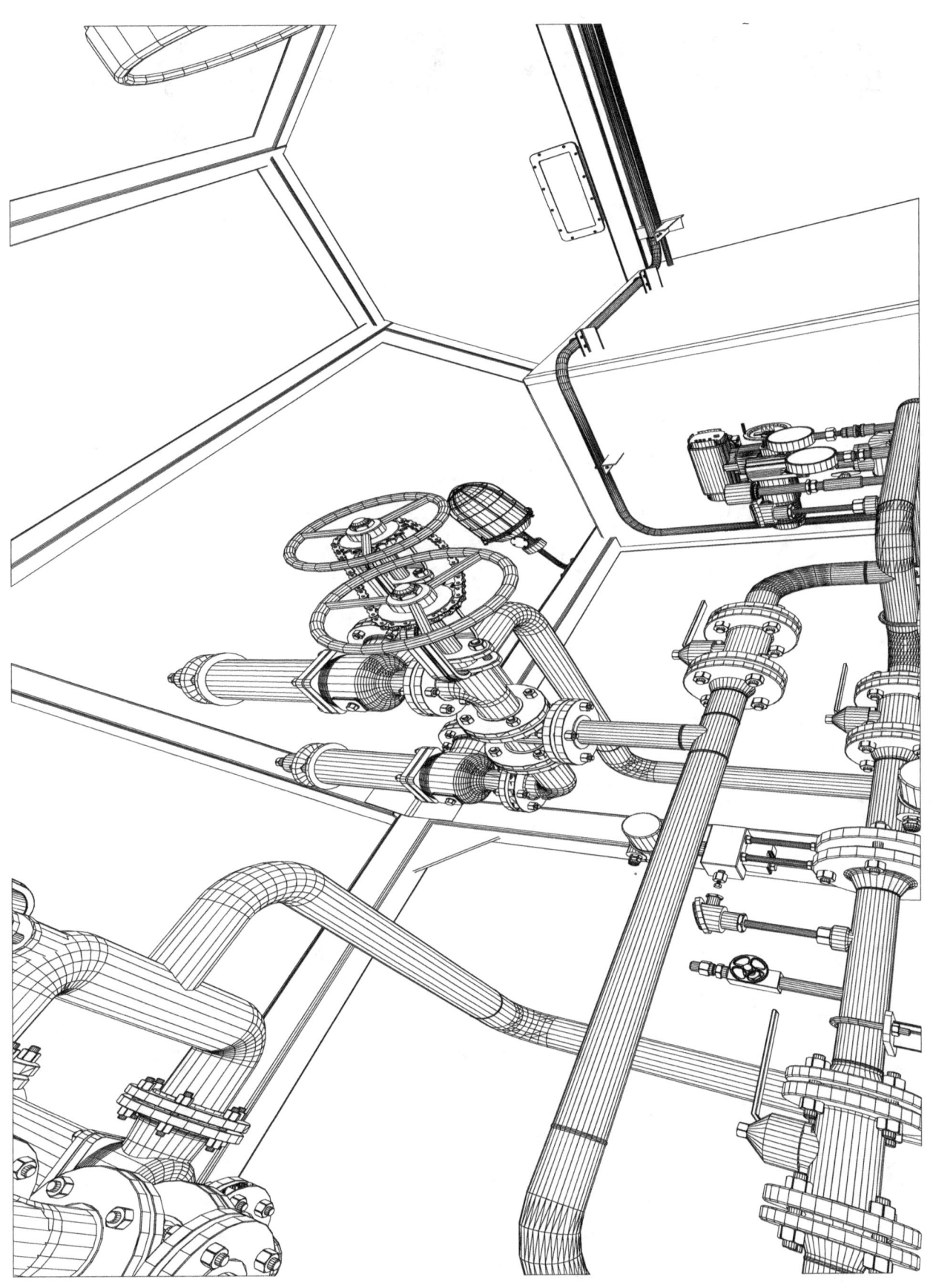

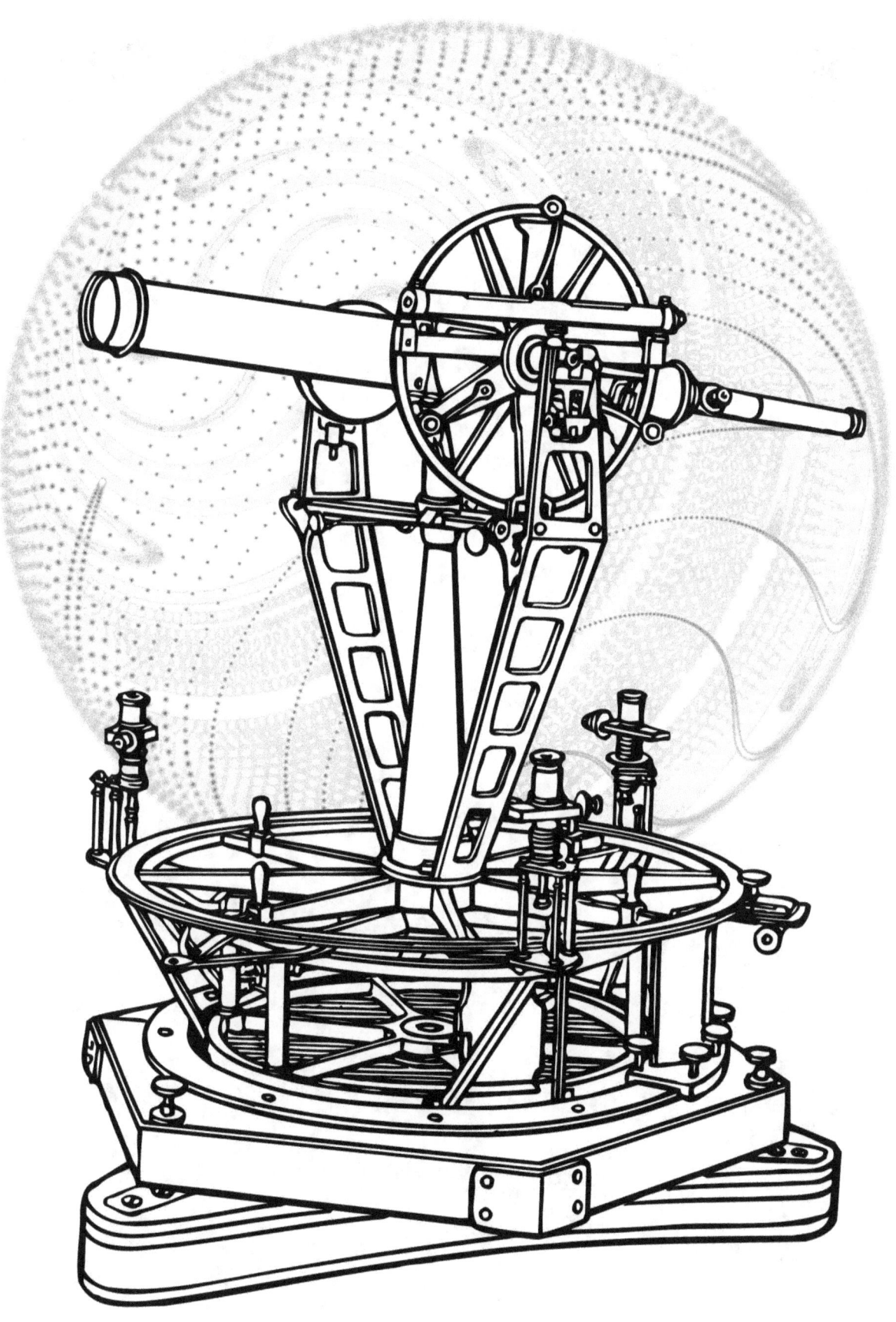

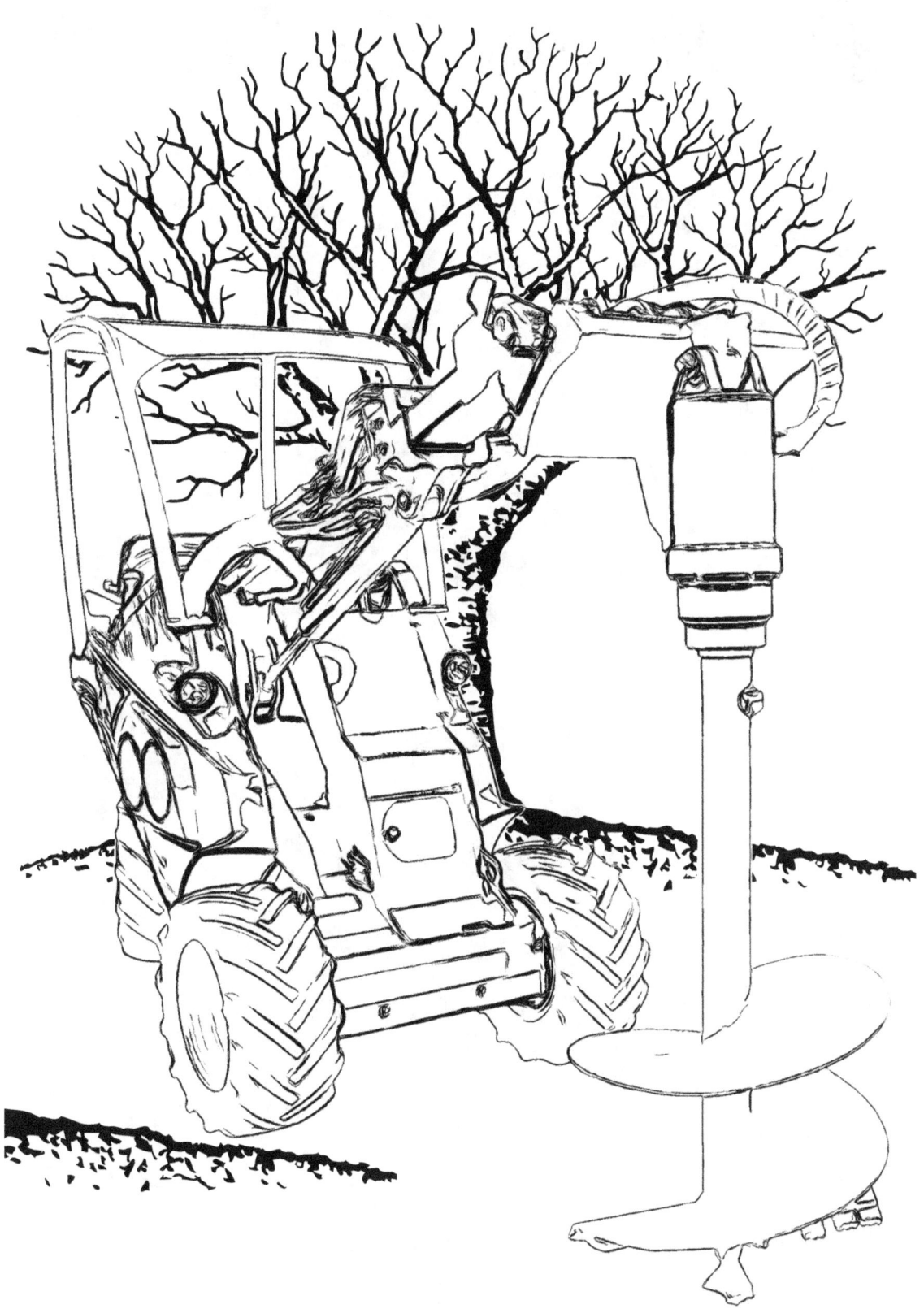

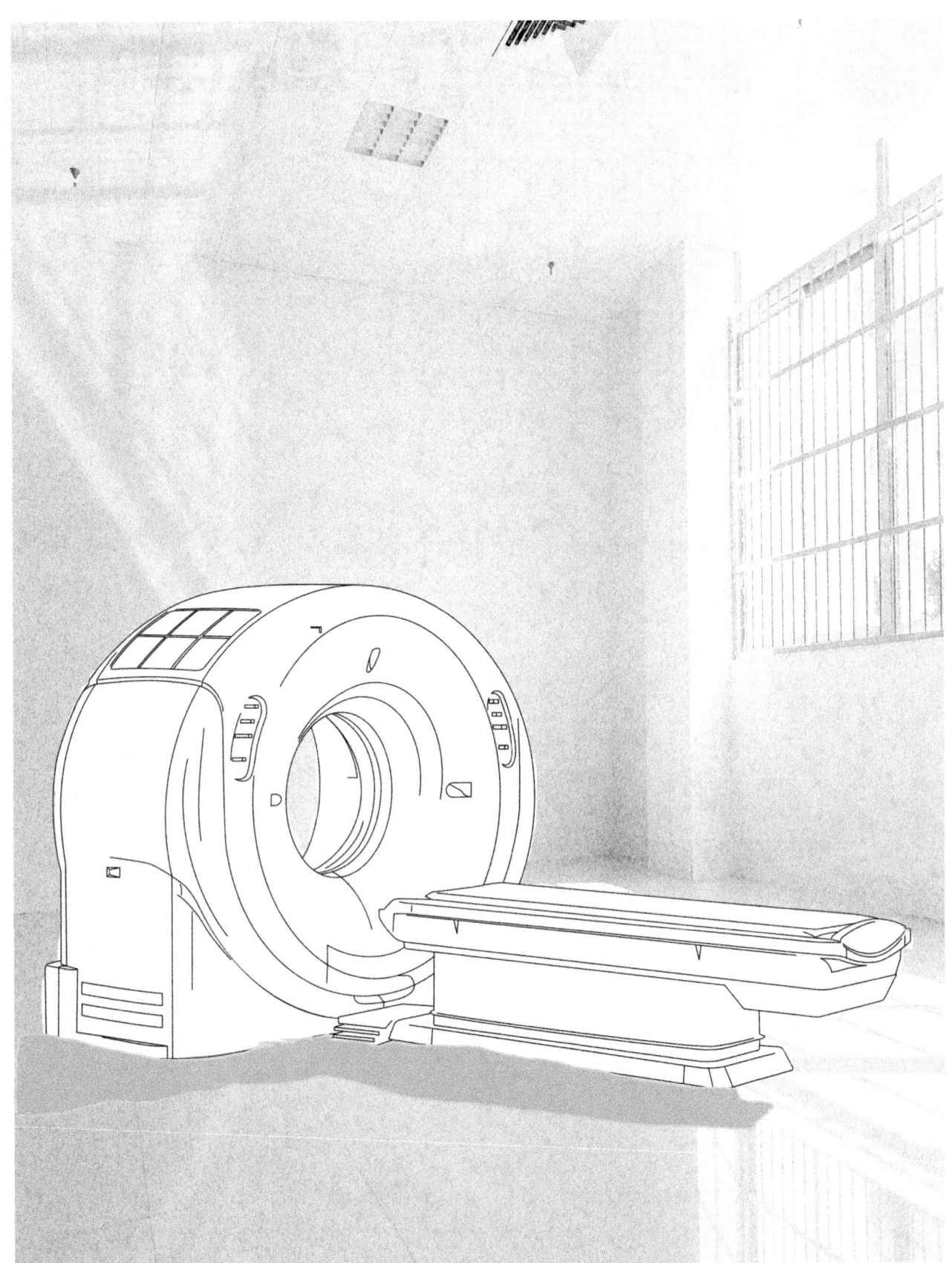

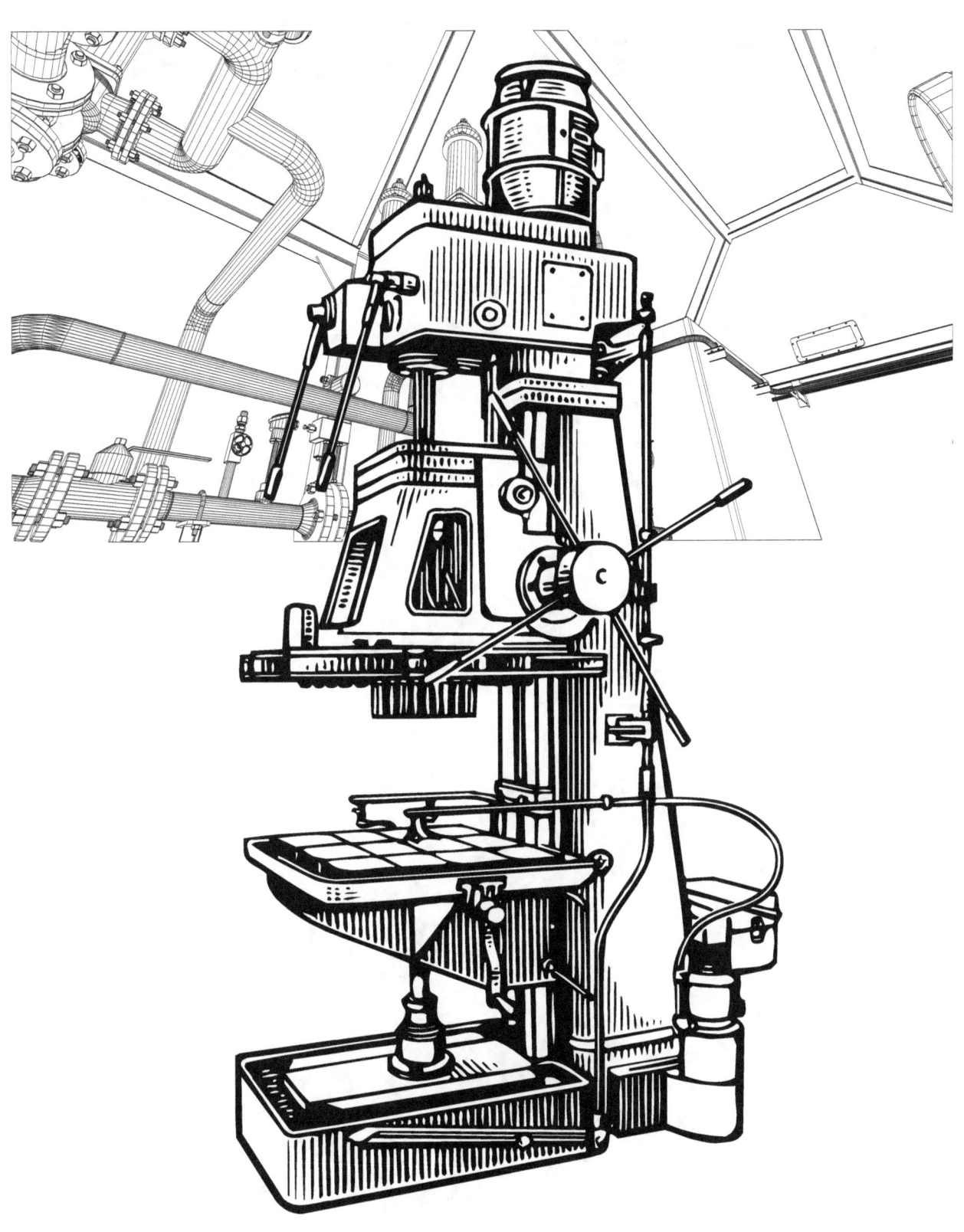

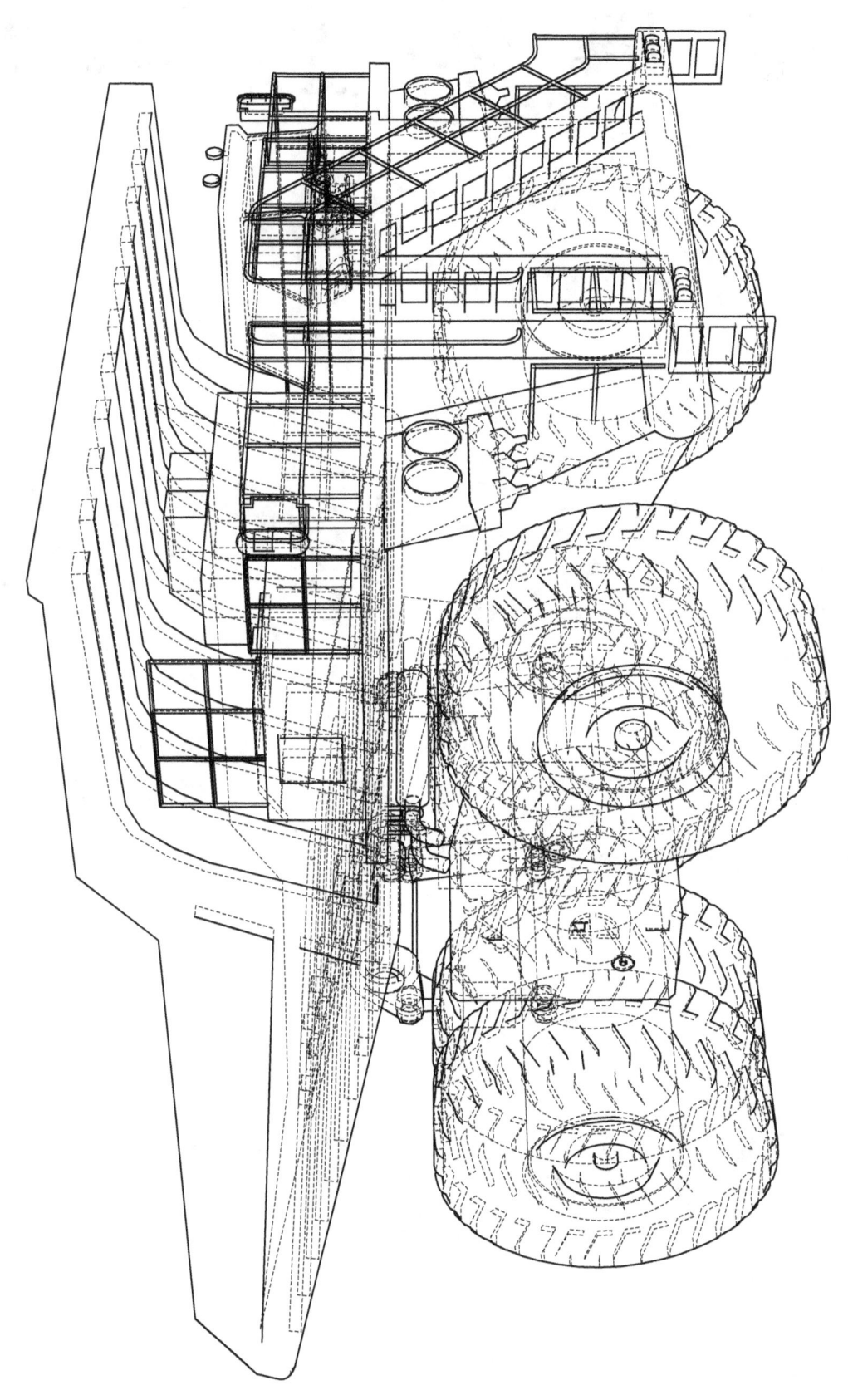

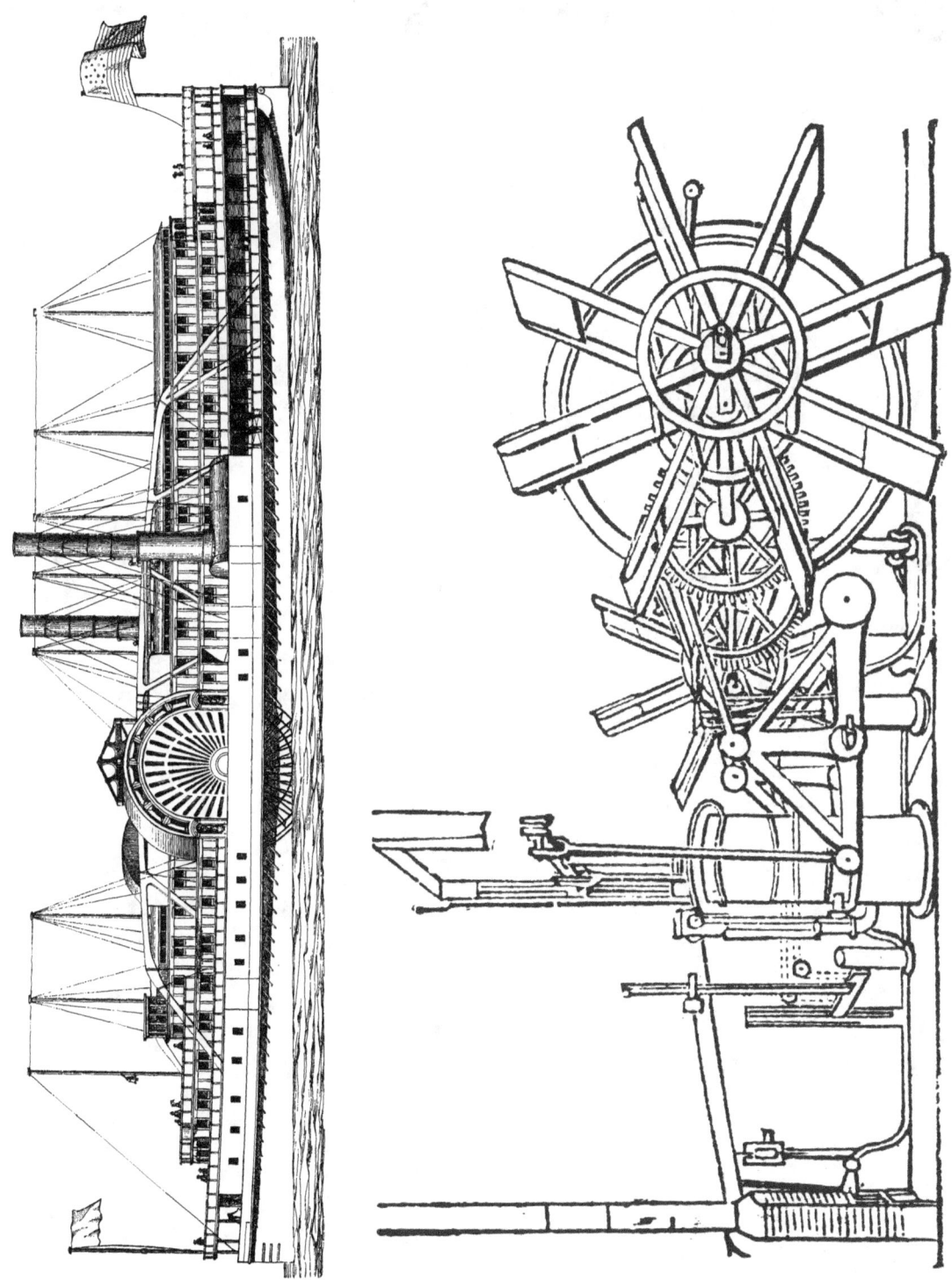

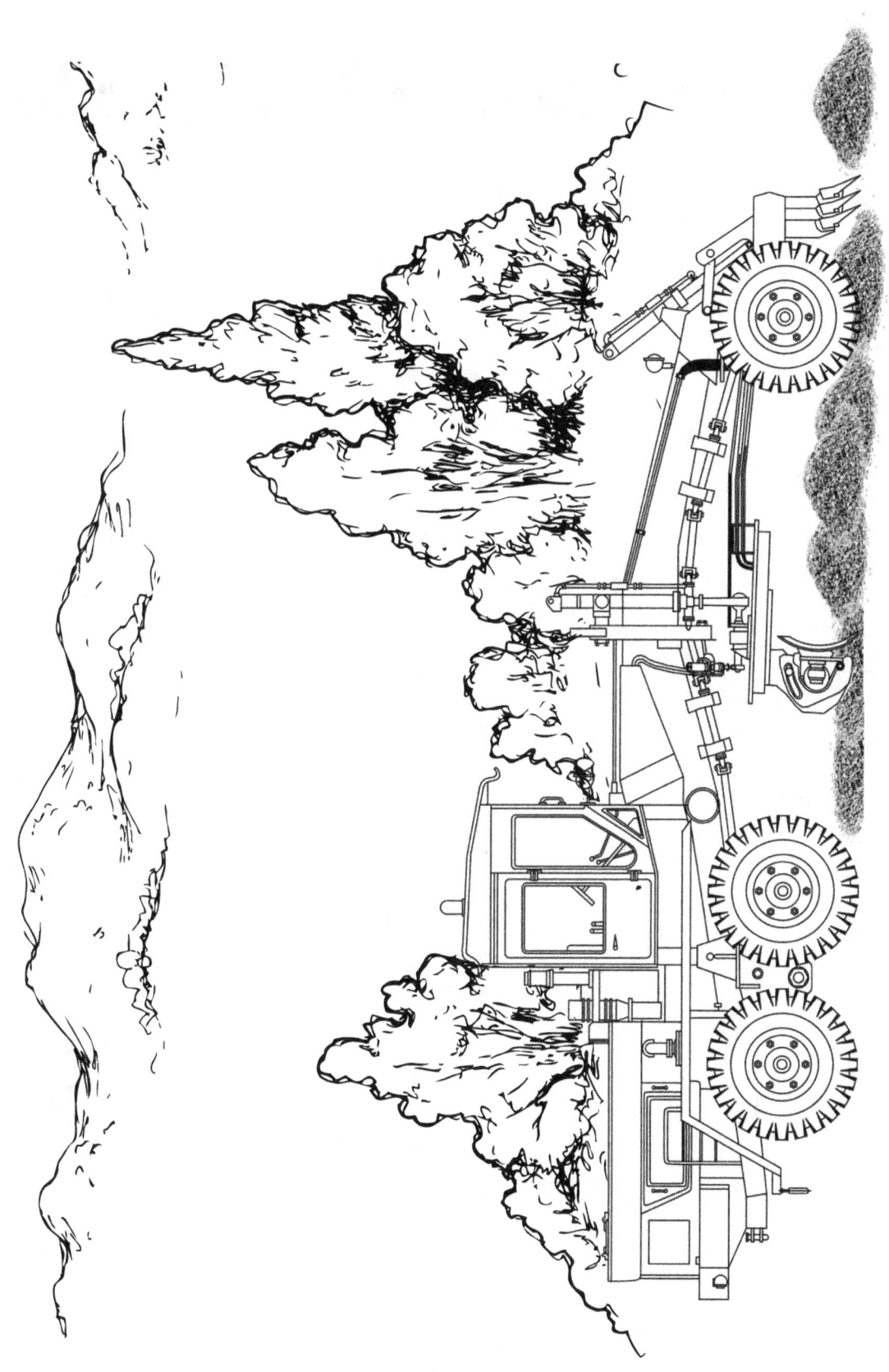

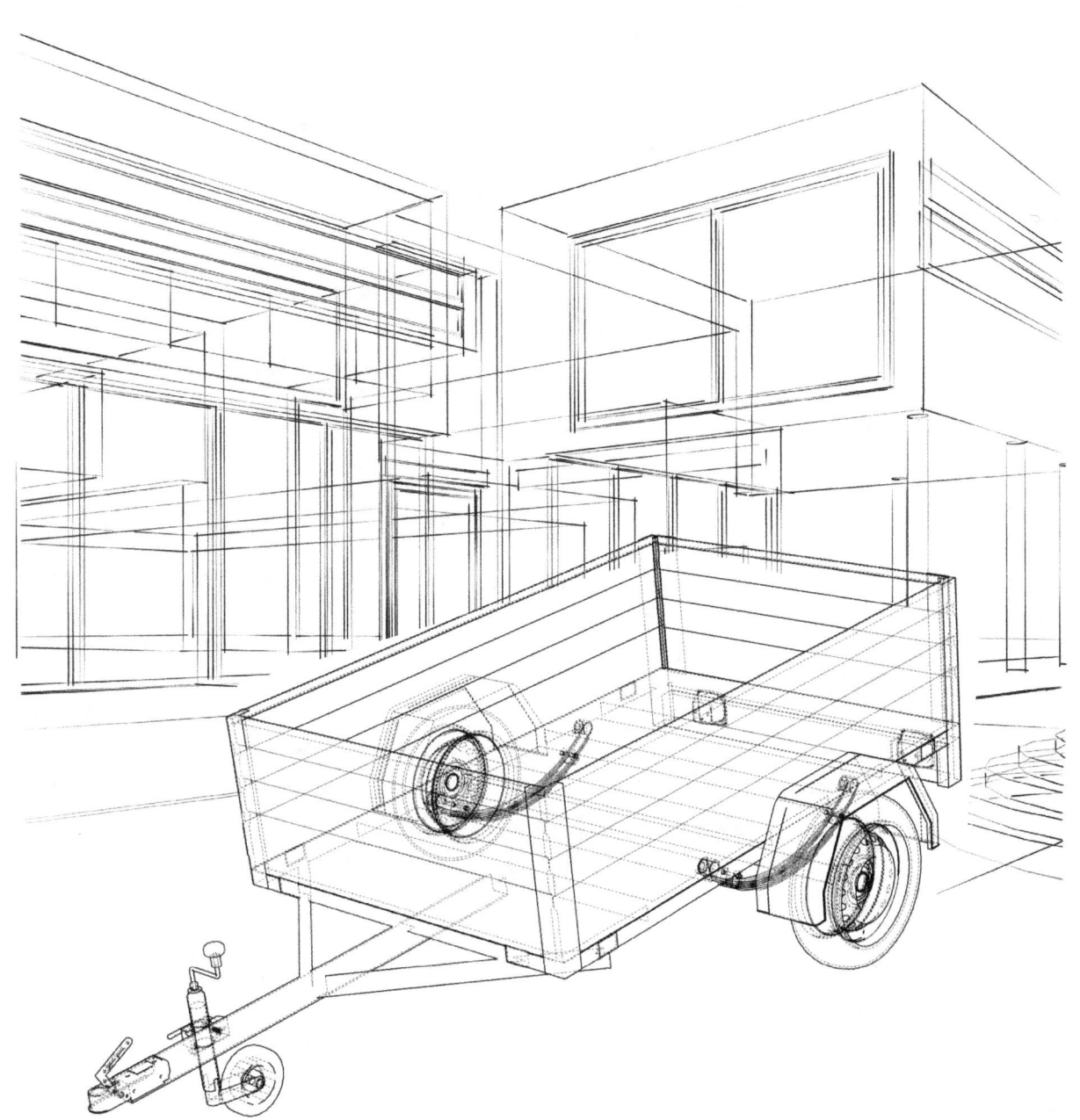

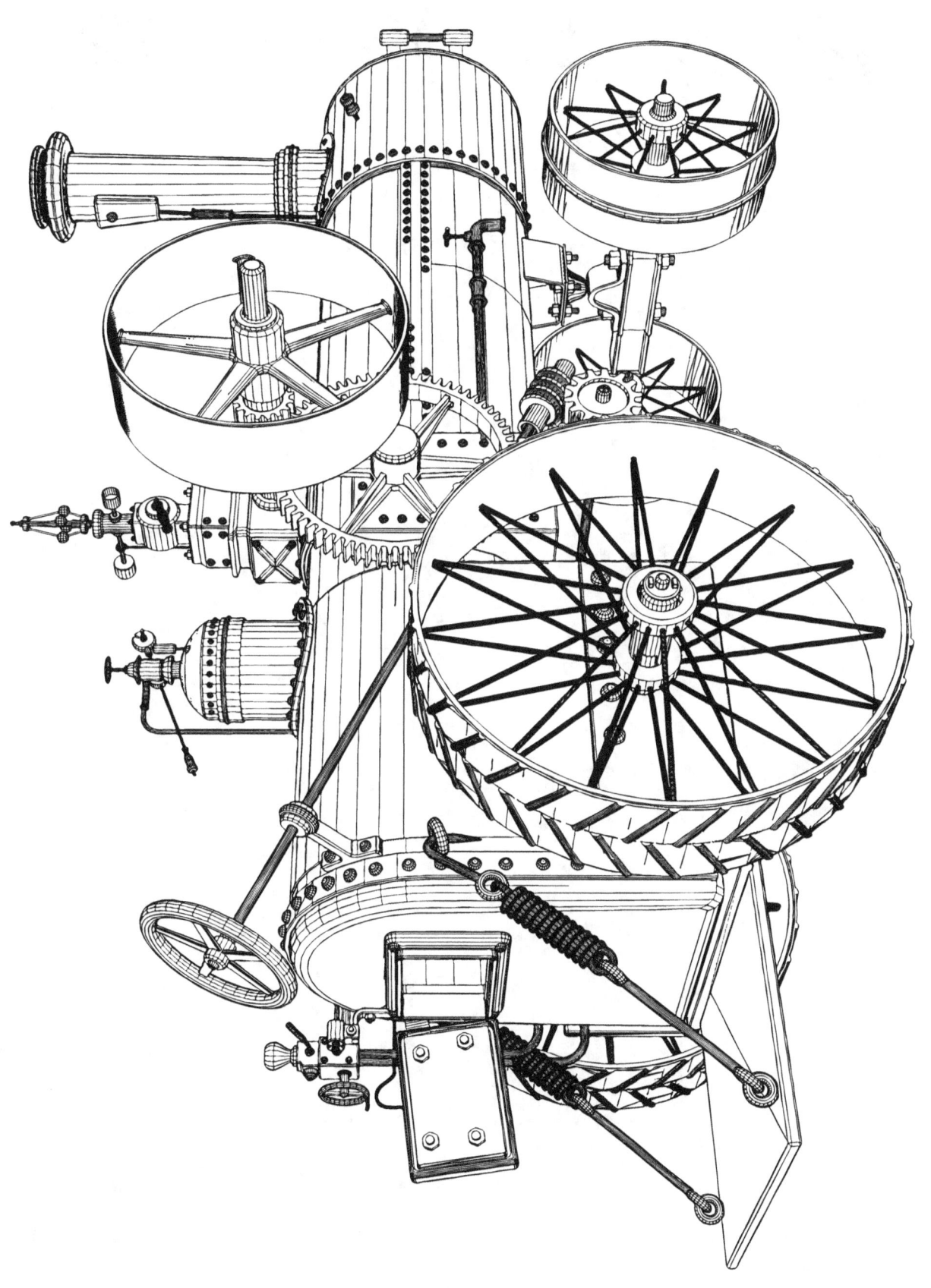

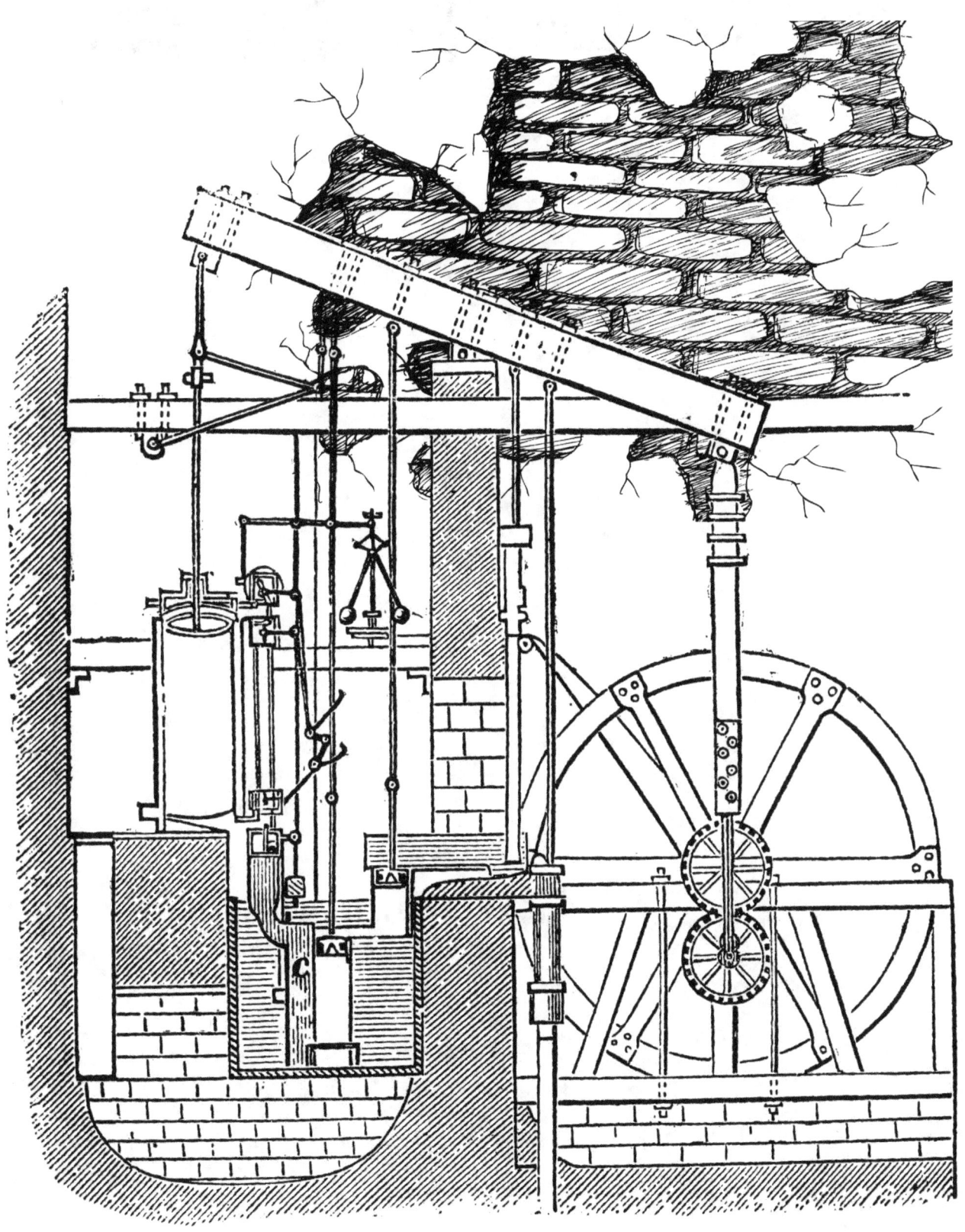

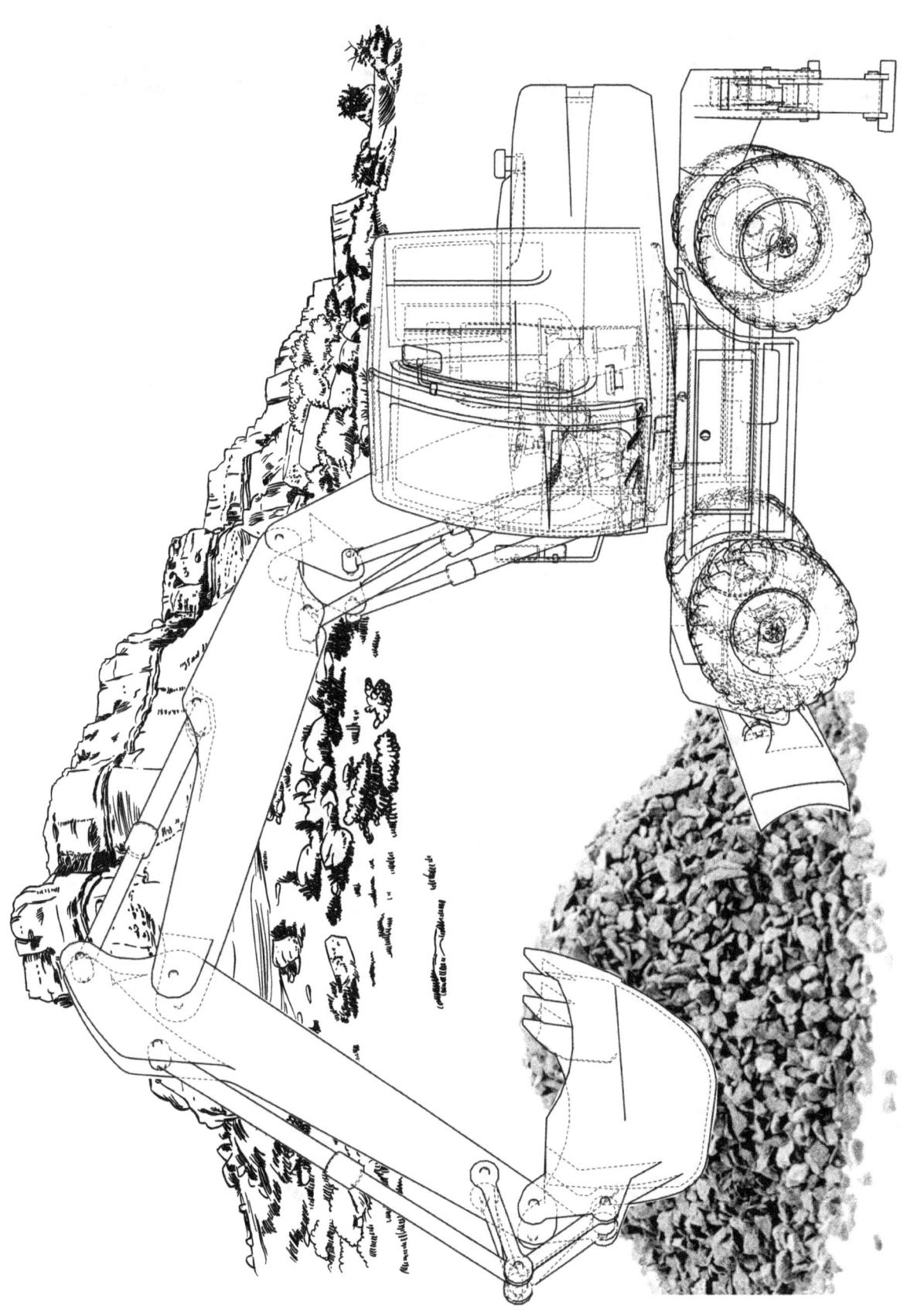

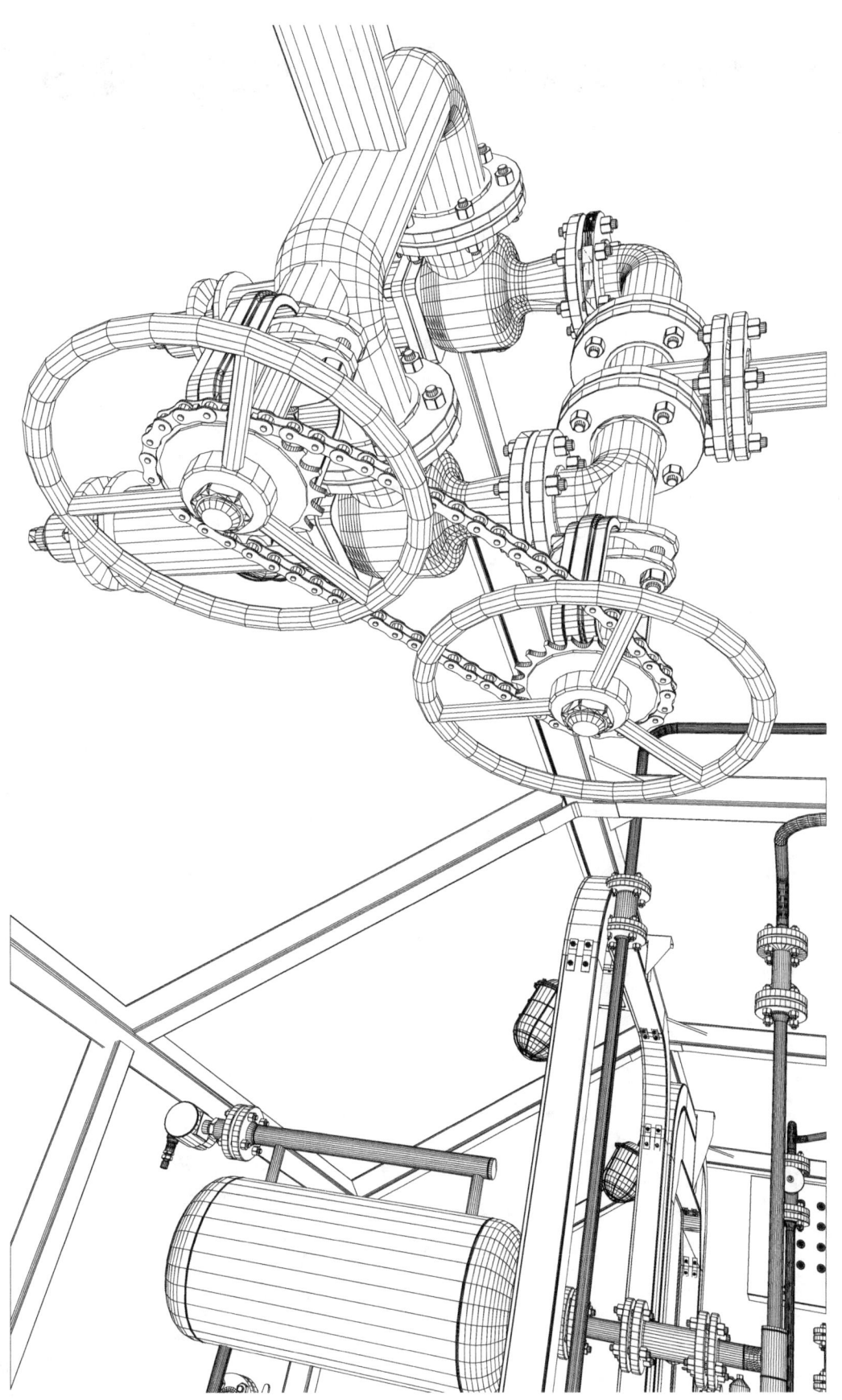

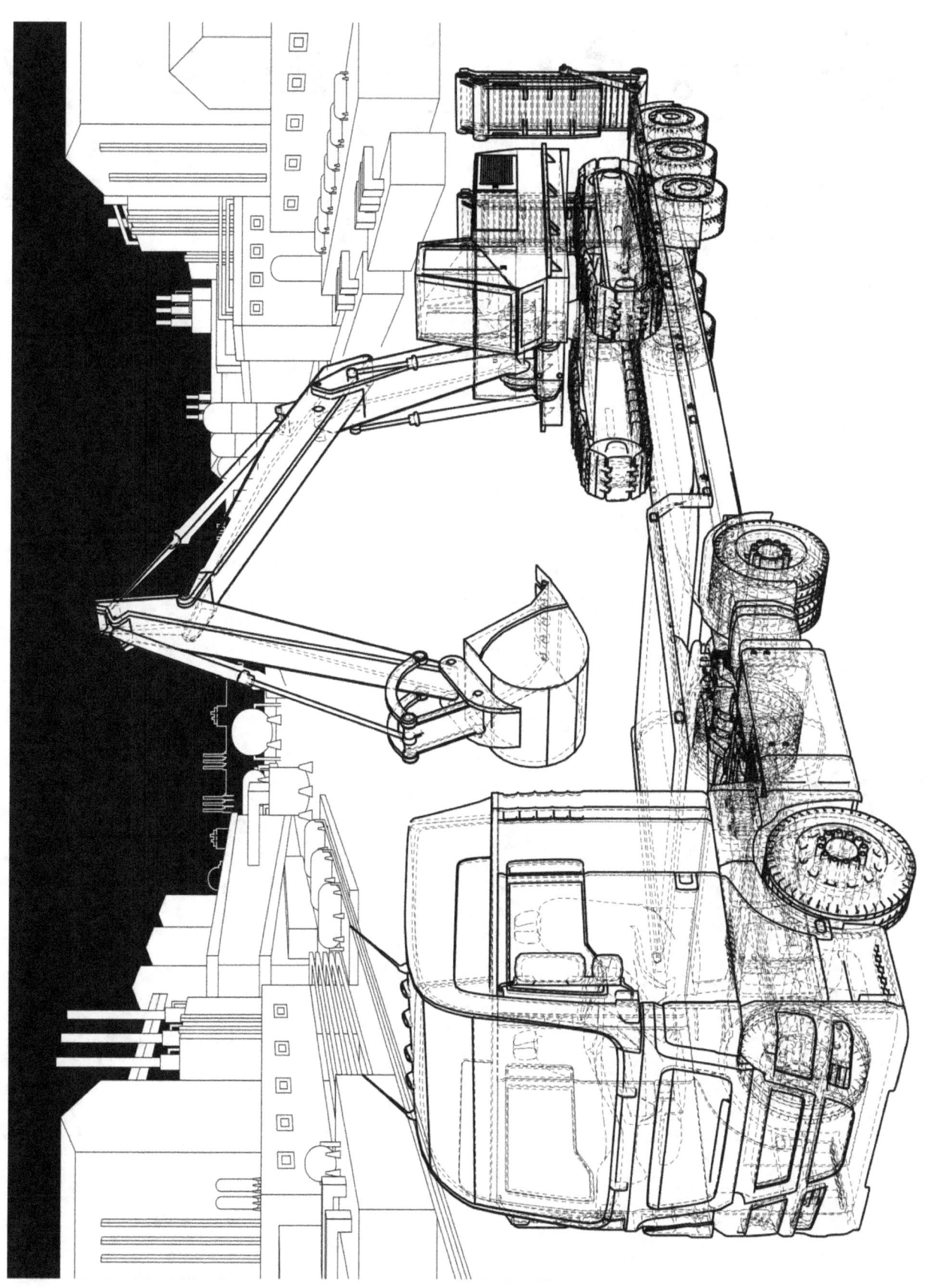

We hope that you enjoyed our Machinery and Equipment Coloring Book. If you liked these machines, maybe you would like our Ships Coloring Book as well.

A SELECTION of RECOMMENDED ART SUPPLIES

Crayola
Fine Line Markers – 40 Count

Brush Markers – Dual Tip w/Ultra Fine Marker – 36 colors 16 Markers

Tanmit
60 Colors 'Calligraphy Brush Marker Pens' – Dual Tip Pastel Bullet and Fine Point Blending Markers

Ohuhu
Watercolor Brush Markers – 20 Color

Qianshan
Colored Pencils Pencil Case – Holds 202 colored Pencils or 136 Gel Pens

www.ingramcontent.com/pod-product-compliance
Lightning Source LLC
Chambersburg PA
CBHW081451220526
45466CB00008B/2599